Patrick KADJAT A TSHEY

Standards and Regulations in Geology and Mining

Patrick KADJAT A TSHEY

Standards and Regulations in Geology and Mining

Challenges, history and prospects for the DRC

ScienciaScripts

Cover image: www.ingimage.com

This book is a translation from the original published under ISBN 978-620-6-70960-2.

Publisher:
Sciencia Scripts
is a trademark of
Dodo Books Indian Ocean Ltd. and OmniScriptum S.R.L publishing group

120 High Road, East Finchley, London, N2 9ED, United Kingdom
Str. Armeneasca 28/1, office 1, Chisinau MD-2012, Republic of Moldova, Europe
Managing Directors: Ieva Konstantinova, Victoria Ursu
info@omniscriptum.com

Printed at: see last page
ISBN: 978-620-8-64452-9

FOREWORD

Against a backdrop of increasing pressure on natural resources and heightened environmental and social challenges, it is essential to explore the standards and regulations governing the mining industry, particularly in the Democratic Republic of Congo, which is rich in mineral resources but faces complex challenges.

This book serves as a tool for understanding the evolution of standards in this sector, from rudimentary standards to modern regulations. The reader is encouraged to reflect on the fundamental role of standards in science, human relations and the evolution of industry.

It was emphasised that the implementation of standards is a strategic lever for a sustainable future. The proposal for a Congolese Mining Reporting Standard (NCRM) embodies the commitment to scientific transparency and the integration of environmental and community concerns.

By adopting a multidisciplinary approach, this book demonstrates that properly applied standards can transform the mining industry into an engine for development, benefiting the country and local communities in particular. It calls for dialogue between government, business and civil society to ensure that resources are exploited responsibly.

In short, this book contributes to the debate on standards in geology and the mining industry, inviting readers to consider sustainable practices that benefit everyone.

*Mr **KIANDA KIBAWA Emmanuel***

First Vice-Chairman of the Special Commission of the National Order of Geologists of the Democratic Republic of the Congo and Chief Director of the Mining Department at the National Institute of Geologists of the Democratic Republic of the Congo.

DRC National Ministry of Mines.

TABLE OF CONTENTS

ACKNOWLEDGEMENTS

I would like to express my sincere thanks to Mr Jean-Felix MUPANDE KAPWA, Chairman of the Special Commission of the Ordre National des Géologues de la République Démocratique du Congo (OGRDC), and to Dr Yvon NDALA TSHIWISA, Permanent Secretary of the same Commission, for their support.

I would also like to extend my warmest thanks to my wife, Mrs Jynette MWAMBA TSHEY, for her unfailing support in the preparation of this book.

Finally, I would like to express my deep gratitude to Mr Emmanuel KIANDA KIBAWA, whose preface enriches this work with his enlightened perspective and generous support.

INTRODUCTION

In a world where natural resources are both a precious asset and a potential source of conflict, the need responsible and ethical management of exploration and mining activities has never been more pressing. Standards, whether social, environmental or technical, are crucial elements in providing a framework for these activities and ensuring that they are carried out in a sustainable manner. This book explores the subject of standards in geology and mining, highlighting their fundamental role in regulation, worker protection, environmental preservation and socio-economic development.

Through a rich analysis of the historical evolution of standards, their application at both international and national levels, and standards specific to the mining industry, we examine how these prescriptions shape practices and influence perceptions of the sector. From the importance of ISO standards to the challenges posed by transparency in the declaration of mineral resources, each chapter aims to shed light on the many facets of a complex standards system.

Focusing on the Democratic Republic of Congo, a country rich in resources but facing significant challenges, we will highlight the urgent need to harmonise mining declaration and reporting standards. We will analyse how the establishment of a normative framework, supported by the Ordre National des Géologues de la République Démocratique du Congo (OGRDC), such the Norme Congolaise de Reporting Minier (NCRM), could boost investor confidence and guarantee ethical and responsible resource management.

This book argues for an integrated approach that recognises the importance of standards as strategic levers to support sustainable and responsible mining. Far from being mere regulatory constraints, they offer opportunities to promote sustainability, social responsibility and integrity in the industry, while contributing to the socio-economic prosperity of local communities. In this exploration, we invite the reader to examine the complex issues surrounding mining regulation and to consider innovative solutions to current challenges.

CHAPTER 1

STANDARD

The notion of a standard is a rich and multifaceted concept that is crucial to the regulation of human interaction, the organisation of behaviour and the definition of quality standards, whether in the social, legal, technical or scientific fields. A norm can be defined as a rule, criterion or standard that regulates behaviour, actions or products within a society or specific field. It serves as a reference for assessing what is considered acceptable, desirable or compliant for a given group. As a frame of , norms not only determine expectations of personal and professional conduct, but also support the structure and cohesion of a community.

1.1. Types of Standards

1.1.1.Standards Social

Social norms encompass the implicit or explicit rules that guide the behaviour of individuals in a given society. They are often deeply rooted in culture, traditions and shared values, and regulate a variety of activities ranging from lifestyle to communication and interpersonal relationships. Examples include polite behaviour, respect for elders and dress codes. Non-compliance with these norms can result in social stigma, ranging from passive disapproval to active sanctions such as exclusion or rejection.

1.1.2.Standards Legal

Legal standards, on the other hand, are rules established by competent authorities such as governments, local authorities and other official institutions. They govern the behaviour of citizens and institutions through laws, regulations and decrees. These standards are often the result of rigorous legislative processes, in which public discussion and legal expertise play a crucial role. Failure to comply with legal standards can result in significant consequences, such as fines, imprisonment or other legal sanctions, underlining their importance in maintaining public order.

1.1.3.Standards Techniques

Technical standards are specifications drawn up to define safety, quality and efficiency requirements for products or processes. These standards guarantee the necessary compatibility between different systems and ensure a minimum level of quality. For example, ISO (International Organisation for Standardisation) standards establish international standards in various fields, while CE (Conformité Européenne) standards guarantee that products comply with European Union safety requirements. Implementing these standards is often essential to guarantee consumer confidence and user safety.

1.1.4.Standards Scientific

In science, standards are criteria established to guarantee the rigour of research, experimentation and the validation of results.They are fundamental for ensuring reproducibility and credibility results science. Rigorous experimental protocols, ethical research standards and laboratory health and safety guidelines concrete examples. For a study to be validated by its peers, researchers often have to comply with strict standards, which reinforces the credibility and integrity of scientific research.

1.2. Features of Standards

Standards have several essential characteristics that make them both fundamental and complex. Firstly, their institutionalised nature means that they are often integrated into the very structure of a society or organisation, becoming constant points of reference in day-to-day operations. Furthermore, failure to comply with standards often leads to a variety of sanctions, ranging from social disapproval in the case of social standards to legal proceedings in the case of legal standards.

Standards are not static; they evolve over time, influenced by cultural, social and technological changes. This dynamism enables them to adapt to new realities and societal challenges. Finally, some standards are universal, such as human rights, while others are contextual or specific to a given group. This variability calls for a critical analysis to understand the extent to which they need to be applied in different contexts.

1.3. Establishment of Standards

Standards can be established by various means. In the case of social and technical standards, consensus between stakeholders, such as civil society and the professionals concerned, is often the basis for their acceptance. This involves in-depth discussions, negotiations and sometimes the compromises needed to find common ground.

For legal standards, formal processes such as legislative sessions, decrees or referendums are used to establish them, often accompanied by public debate and legal expertise. In the scientific field, standards are mainly established through rigorous research practices and peer validation. This process ensures that the standards adopted are based on solid empirical foundations, which is essential for the credibility and integrity of scientific standards.

1.4. Issues associated with Standards

The concept of standards raises significant issues and debates. On the one hand, there is a tension between flexibility and rigidity: some people argue that overly strict standards can innovation and creativity, while others believe that they are essential for maintaining order, security and respect within society. On the other hand, the question of cultural versus universal standards is the subject of lively debate, particularly in the field of rights. This debate raises questions not only cultural integrity and diversity, but also about respect for individuals and communities in the implementation of these standards. In short, the notion of a standard is a fundamental concept that regulates and structures human interaction in a variety of fields. Whether social, legal, technical or scientific, standards play a crucial role in establishing standards of conduct and quality, thereby guaranteeing cohesion and order within societies. Their diversity and complexity reflect the richness of social dynamics and the contemporary challenges facing individuals and institutions.

As dynamic elements, standards are not static; they evolve in response to cultural, technological and social changes. This adaptability underlines the importance of critical analysis and ongoing dialogue in understanding and applying these rules in a variety of contexts. The issues associated with standards, in particular the tension between flexibility and rigidity, and the debate about universality versus cultural specificity, call for in-depth reflection on how we define and comply with these standards.

Thus, whether through the prism of social, legal, technical or scientific standards, it is imperative to recognise and appreciate their impact on our daily lives and on the way societies function. Ultimately, standards, while being instruments of regulation, must also serve as catalysts for positive change and progress, respecting the diversity and richness of human experience.

CHAPTER 2

HISTORY STANDARDS IN MINING EXPLORATION AND GEOLOGY

The history of standards in geological exploration and mining geology is deeply rooted in the evolution of human societies. As an essential activity, the exploration and exploitation of natural resources have not only shaped civilisations, but have also had considerable social, economic and environmental implications. This chapter traces the historical development of the standards that have guided these practices, while highlighting the scientific and societal developments that have influenced their formulation.

2.1. Prehistoric Origins

2.1.1. Discoveries Primitives

The first attempts at geological exploration date back to prehistoric times. Human societies discovered minerals and precious metals such as copper, gold and silver, which they used to make tools, weapons and ornaments. These resources were essential for the survival and development of the first civilisations. Extraction techniques were rudimentary, based on empirical knowledge and traditional practices.

2.1.2. Towards Systematic Practice

With the rise of civilisations, particularly in Mesopotamia, Egypt and the Indus Valley, exploration practices began to take shape. The ancient Egyptians, for example, developed more sophisticated extraction for precious stones and metals, laying the foundations for more systematic exploration methods. This period marked the beginning of more organised interaction between man and his geological environment.

2.2. The Evolution of Geology as Science

2.2.1. The first classification systems

Over the centuries, geology has gradually evolved into a science. The work of thinkers such as Aristotle and later geologists such as Georges-Louis Leclerc, Comte de Buffon, contributed to our understanding of geological formations. The first geological maps, drawn up in the

18th century, provided visual tools for exploring natural resources, paving the way for more methodical practices.

2.2.2. The Industrial Revolution and its implications

The Industrial Revolution of the 19th century radically changed mining practices. The development of machinery enabled large-scale, more efficient extraction. However, this also led to problems with worker safety and environmental degradation. The consequences of these practices have given rise to growing concern within companies, leading to calls for regulation.

2.3. The Recognition of Standards in the Years 1960

2.3.1. The emergence of safety standards

The importance of mining geology standards began to take shape in the 1960s. Conflicts over resource management, exacerbated by tragic mining accidents, highlighted the need for safety standards to protect workers. Organisations such as the International Labour Organisation (ILO) began to draw up recommendations to improve working conditions in the mining sector.

2.3.2. Corporate Social Responsibility

At the same time, awareness of the social impacts of mining activities has led to growing recognition of corporate social responsibility. Mining companies began to be held accountable for the effects of their operations on the surrounding communities, paving the way for ethical and sustainability standards.

2.4. The emergence of environmental standards

2.4.1. The 1980s and 1990s: Ecological awareness

Environmental concerns intensified in the 1980s and 1990s. Reports of the negative impacts of mining on ecosystems and local communities led to global mobilisation. Conferences such as the Rio Declaration in 1992 formalised international commitments to sustainable mining practices.

2.4.2. Adoption of environmental standards

This period also saw the emergence of specific environmental standards. Initiatives such as ISO 14001, which promotes environmental management systems, were adopted by many mining companies, marking a step towards responsible resource management.

2.5. The Contemporary Era: Globalisation and International Standards

2.5.1. Harmonisation of practices

With globalisation, mining markets have become increasingly interconnected, making it necessary to harmonise standards. Organisations such as the World Miners' Commission (WMC) and the International Association Geological Society (IAG) have played a crucial role developing guidelines to ensure responsible, ethical and sustainable geological exploration.

2.5.2. Modern sustainability initiatives

Today, initiatives such as the Extractive Industries Transparency Initiative (EITI/ITIE) and the Responsible Mining Initiative (RMI) demonstrate the sector's move towards more sustainable and transparent practices. These standards aim to meet the growing expectations of stakeholders in terms of sustainability and social responsibility.

This chapter has therefore traced the historical development of standards in geological exploration and mining geology, revealing how these essential practices have been shaped by social, economic and environmental dynamics over time. From From prehistoric times, when exploration was based on rudimentary methods, to the emergence of more systematic structures in ancient civilisations, we have seen a steady progression towards more elaborate extraction .The evolution of geology as a science has also played a very important role, particularly with the Industrial Revolution, which introduced both significant opportunities and challenges in terms of safety and environmental impact. The growing recognition of safety standards and corporate social responsibility in the 1960s marked a turning point, highlighting the need to protect not only workers, but also the communities affected by mining activities. The 1980s and 1990s were emblematic of growing environmental awareness, leading to the adoption of environmental standards and international commitments to promote sustainable operating practices. Today, in an era of globalisation, the harmonisation of practices and the emergence of modern initiatives demonstrate a growing commitment to sustainability and transparency in the sector. In short, the history of standards in geological exploration and mining geology is a reflection of changing societal values and environmental challenges. This history highlights the importance of an evolving regulatory framework that is capable of adapting to contemporary issues while guaranteeing responsible and ethical mining. natural resources. The future of this sector will require continued vigilance and a strong commitment to integrating sustainable and responsible practices at the heart of exploration and mining.

CHAPTER 3

THE IMPORTANCE OF STANDARDS IN THE MINING INDUSTRY

As we saw earlier, standards play a fundamental role in the mining industry, acting essential guidelines to ensure safe, efficient and sustainable operations. They have far-reaching implications that go beyond the simple extraction of resources, touching on quality, safety, the environment and social responsibility.

3.1. Guarantee the quality of Resources

Applying rigorous standards is crucial to ensuring the quality of the minerals and resources extracted. This includes selection , exploration and extraction methods, and processing techniques that meet specific criteria. Companies that comply with these standards are in a better position to offer high-quality products that meet the requirements of international markets. This not only increases competitiveness, but also maximises the value of mineral resources.

3.2. Protection of Workers

Worker safety is a major concern in the mining sector. Standards define practices and protocols that minimise the risk of accidents and injuries. This includes mandatory training on mining risks, the use personal protective equipment (PPE), as well as well-established evacuation and emergency procedures. By complying with these standards, companies not only protect their staff, but also reduce the costs associated with accidents, occupational illness and proceedings.

3.3. Protecting the environment

The mining industry has often been criticised for its environmental impact. Environmental standards ensure that companies implement responsible practices to reduce and mitigate the harmful effects of their operations. This includes the management of mining waste, the restoration of sites after exploitation, and the use of less polluting technologies. By adopting these measures, mining companies can reduce their ecological footprint and contribute to the sustainability of natural resources.

3.4. Reducing Costly Disputes

Failure to comply with standards can lead to costly litigation and legal penalties. Companies that invest in standards compliance minimise the risk of legal action, penalties and disruption to their operations. This robust regulatory framework not only protects businesses, but also helps build trust among stakeholders, including governments and local communities.

3.5. Improve your reputation

Companies that meet high standards enjoy a better reputation in the marketplace. A strong reputation for safety, quality and environmental responsibility can improve relationships with investors, customers and business partners. Transparency and commitment to best practice attract stronger partnerships and loyal customers, increasing the long-term value of the business.

3.6. Facilitating Access to Financing

Standards also play a key role access to finance. Financial institutions and investors often evaluate compliance with standards as an essential criterion before deciding to grant funding. Companies that demonstrate rigorous adherence to standards are perceived as less risky, enabling them to negotiate better financing terms and attract investment. In short, standards are an essential pillar of the mining industry, significantly influencing the quality of resources, the protection of workers, the preservation of the environment and the management of legal risks. Their application not only ensures the safety and sustainability of operations, but also strengthens the competitiveness of companies on the global market. By adopting practices that comply with the standards, mining companies improve their reputation and build trust with their , while access to information. necessary funding for their development. Implementing standards is a winning strategy that benefits the entire mining ecosystem. It reconciles the exploitation of natural resources with social and environmental responsibility, thereby meeting the growing expectations of consumers, investors and local communities. In a world where sustainability issues are becoming increasingly crucial, the importance of standards in the mining industry cannot be underestimated. Adapted and respected, they pave the way for a safer, more responsible and more prosperous future for all players in the sector.

CHAPTER 4

CONVERGENCE OF STANDARDS AND THEIR SCOPE OF APPLICATION

Standards, whether technical, industrial or professional, are essential for structuring human activities. In the mining industry, these standards play a crucial role in laying down a framework that ensures the safety of operations, protects the environment and guarantees the quality of the resources extracted. Although they vary from country to country and region to region, standards share common objectives that promote sustainable and responsible mining development.

4.1. Common aspects of Standards

4.1.1. Standardisation objective

Standardisation is the cornerstone of standards. By establishing clear and measurable criteria, standards aim to standardise practices across various sectors. This not only ensures the quality and safety of products and services, but also communication and trade on a global scale. In the mining industry, for example, standardising methods for extracting and processing minerals can lead to more predictable results and lower operating costs. In addition, this standardisation helps to minimise errors and improve productivity, by providing companies with a frame of reference for developing their processes.

4.1.2. Consensus

standards development process often based on consensus between several stakeholders. These include experts from the sector, representatives of government bodies, industry, as well as consumers and NGOs. This consensus model ensures that standards are balanced and take into account the varied interests of the stakeholders involved. For the mining industry, this means that concerns about worker safety, environmental protection and economic profitability are all taken into account, enhancing the acceptability and effectiveness of standards.

4.1.3. Voluntary or mandatory

Standards can be categorised into two main types: those adopted on a voluntary basis and those imposed by legislation. Voluntary standards, such as certain ISO certifications, offer companies the opportunity to stand out in the marketplace. On the other hand, mandatory standards, often by governments, are essential to ensure compliance with minimum safety, quality and environmental requirements. In the mining sector, a combination of voluntary and mandatory standards can encourage companies to go beyond regulatory requirements, fostering a culture of safety and sustainability.

4.1.4. Documentation

Documentation of standards is crucial to their effective implementation. These documents, which may include manuals, guides and detailed procedures, specify technical requirements, test methods and assessment . For the mining industry, clear documentation helps to ensure that everyone involved understands the expectations and obligations relating to safety, quality and environmental management. Training manuals can also be developed to ensure that workers and managers are up to date with the latest practices and procedures.

4.1.5. Evolution

Standards must evolve to remain relevant in the face of technological advances and new practices. This includes the periodic revision of existing standards to incorporate scientific discoveries, technological innovations and feedback from the field. In the mining industry, for example, the emergence of new extraction techniques or monitoring technologies can prompt updating of standards to ensure that methods remain safe and effective. This ability to evolve is essential to maintaining the competitiveness and sustainability of mining operations.

4.1.6. Impact on Competitiveness

The adoption of quality and safety standards can have a significant impact on the competitiveness of companies in national and international markets. Companies that comply with high standards are often perceived as more reliable and rigorous, which can enable them attract more customers and business partners. In addition, compliance with international standards can facilitate access new markets, strengthen brand reputation and, ultimately, contribute to economic growth. In mining industry, compliance with standards can also attract investment, as investors are increasingly committed to social and environmental responsibility.

4.2. Scope Application of Standards

4.2.1. Industry and Commerce

Standards are ubiquitous in industry and commerce. In the mining industry, standards ensure that manufactured products, such as mining and processing equipment, meet strict quality and safety criteria. In this case, standards such as ISO 9001 define criteria for quality management, enabling companies to ensure that their operations are aligned with performance requirements. This reduces the risk of equipment failure, improving operational safety.

4.2.2. Services

Standards are not limited to products; they also extend to services. In the mining services sector, standards such as ISO 9001 consistent quality, whether in employee training, operations management or technical support. This is particularly important in an environment where safety and reliability are paramount. Implementing service standards helps to set clear expectations and improve customer satisfaction.

4.2.3. Environment

Environmental protection is a major issue in the mining industry. Standards such as ISO 14001, which focus on environmental management, enable companies to adopt sustainable practices. These standards encourage assessing the environmental impact of operations, setting up waste management systems and optimising resource consumption. By incorporating these standards, mining companies can reduce their environmental footprint and meet the growing expectations of stakeholders in terms of sustainability.

4.2.4. Health and Safety

Worker health and safety are top priorities in the mining industry. Specific standards, such as ISO 45001, are in place to ensure a safe working environment. They set out requirements for identifying hazards, assessing risks and implementing preventive measures. encourage employee training in safety practices, thereby reducing the number of workplace accidents and promoting a safety culture within mining operations.

4.2.5. Information Technology

With the advent of digital technologies, technical standards have become essential to ensure the interoperability and security of IT systems. In the mining industry, this includes standards for data management, cyber security and the protection of control systems. For example, standards such as ISO 27001 for information security management ensure that data in mining operations is protected against breaches and that systems are resilient in the face of cyber attacks.

4.2.6. International vs. national standards

Standards can be international, such as those developed by ISO, or specific to a country or region. International standards facilitate intercontinental trade and ensure a certain uniformity of practice. However, national standards can also be crucial, as they take into account local socio-economic and environmental contexts. Mining companies must navigate this duality to ensure compliance while optimising their operations.

4.2.7. Interdisciplinary

Standards often touch on several areas and disciplines. In the mining industry, this can include engineering, safety, environmental and social standards. For example, the same mining operation may be subject to safety requirements for workers, natural resource management standards and regulations on the rights of local communities. This interdisciplinarity underlines the importance of considering standards as a whole to ensure an integrated and responsible approach to mining operations.

4.3. Industry standards

4.3.1. Security Operations

Operational safety is a top priority in the mining industry, where the risk of accidents is significant. Standards, such as those established by the Occupational Safety and Health Administration (OSHA) and ISO 45001, set out requirements for for accident prevention and worker protection. This includes protocols for risk assessment, employee training on safety procedures and the use of personal protective equipment. By following these standards, companies can reduce accidents, minimise injuries and create a safer working environment.

4.3.2. Protecting the environment

The environmental impact of mining activities is a major issue. Environmental standards, such as ISO 14001, provide management frameworks to help companies reduce their impact on the environment. This can include environmental impact assessments, waste management procedures and strategies to reduce water consumption. By adopting these standards, mining companies commit to sustainability standards, which are essential for preserving local ecosystems and gaining the approval of surrounding communities.

4.3.3. Quality assurance for Resources

The quality of the resources extracted is crucial to the profitability of mining projects. Standards such as ISO 9001 establish quality management systems that guarantee the traceability of extracted materials and compliance with market requirements. This includes the implementation of quality control procedures at every stage of the process, from extraction to distribution. Compliance with these standards also enhances the reputation of companies, making them more attractive to investors and customers.

4.3.4. Applicability and Durability

The application of standards in mining projects is essential to ensure their long-term sustainability. These standards provide a framework for risk assessment, natural resource management and regulatory compliance.

They also encourage innovation, by encouraging companies to adopt new technologies and more efficient practices. By incorporating rigorous standards, companies demonstrate their commitment to social and environmental responsibility, while promoting economic growth. This translates long-term benefits not only for companies, but also for surrounding communities and ecosystems.

The complexity and diversity of the standards governing the mining industry reflect the importance of establishing solid standards to guarantee the safety, sustainability and quality of operations. This chapter has revealed that, despite geographical variations and sector specificities, there are fundamental commonalities underpinning these standards. Standardisation and consensus are at the heart of this standards landscape. Standardisation enables uniformity of practice that facilitates international trade and guarantees consistent quality, while consensus between various stakeholders ensures that standards are pragmatic and adapted to the realities on the ground. This collaborative framework is essential for creating standards that address both worker safety and environmental concerns.

The dichotomy between voluntary and mandatory standards offers companies strategic flexibility. Voluntary standards, such as ISO, allow companies to differentiate themselves in the marketplace, while mandatory standards ensure the minimum compliance required to protect people and the environment. This coexistence creates a dynamic where companies are encouraged to voluntarily adopt practices that go beyond legal obligations, fostering a culture of safety and sustainability within the sector.

Documentation and the ongoing development of standards also play a major role. Clear documentation facilitates implementation and understanding of requirements, while the evolution of standards ensures that they remain relevant in the face of technological innovation. In a constantly changing sector such as mining, this ability to adapt is essential to maintain competitiveness and meet the growing expectations of stakeholders.

In terms of the scope of application of standards, we have seen that they touch on several crucial aspects such as industry and commerce, services, environmental protection, health and safety, and even information technology. This highlights the need for interdisciplinarity, so that the complex interactions between these different areas can be taken into account to ensure an integrated approach to mining operations.

Implementing standards in the mining industry should not be seen simply as a compliance tool, but as a strategic lever for ensuring sustainability and corporate social responsibility. By adopting rigorous standards, companies can reduce their environmental impact, ensure the safety of their operations, and enhance their reputation with investors and local communities.

In short, standards are an essential basis for structuring the activities of the mining industry. The pursuit of high standards of quality, safety and sustainability is not only a regulatory requirement, but also an economic imperative that conditions the long-term viability of mining projects. Companies that embrace these standards can not only ensure compliance with the rules in force, but also benefit from a competitive position on the international market, while making a positive contribution to the socio-economic development of the regions in which they operate.

CHAPTER 5

INTERNATIONAL STANDARDS

5.1. ISO 14001: Environmental management systems

ISO 14001 is a standard that provides a strategic framework for organisations wishing to manage their environmental responsibilities. It has several key elements:

Improving environmental performance

Companies that adopt this standard commit to integrating sustainable practices into their day-to-day operations. This involves implementing procedures to reduce the ecological footprint, such as using less polluting technologies and promoting environmentally responsible behaviour among employees. This continuous improvement helps to meet the growing expectations of stakeholders, while strengthening the company's reputation in the marketplace.

Efficient use of resources

ISO 14001 encourages companies to optimise the use of their resources, whether water, energy or materials. By reducing waste and adopting more frugal production practices, organisations can make significant savings on their operating costs. In addition, promoting the use of renewable materials contributes to long-term sustainability.

Reducing waste

The standard encourages companies to develop waste management strategies, such as recycling, reusing and recovering materials. These initiatives not only reduce the amount of waste sent to landfill, but also recover valuable resources, creating a more circular and environmentally friendly production cycle.

Compliance with regulations

By incorporating the requirements of ISO 14001, companies gain a framework for complying with environmental legislation, both locally and internationally. This reduces legal risks and strengthens the credibility of the organisation in the eyes of its customers and investors.

5.2. ISO 9001: Quality management systems

ISO 9001 is a global standard that defines the criteria for a quality management system. Its main aspects are as follows:

Meeting customer expectations

Adopting ISO 9001 enables organisations to better understand their customers' needs and adapt their processes accordingly. This includes collecting regular feedback and adjusting products and services to meet market requirements, thereby strengthening customer loyalty.

Improving internal processes

This standard encourages companies to define clear and measurable objectives, monitor performance and commit to a process of continuous improvement. This may involve regular internal audits and staff training to maximise operational efficiency.

Reinforcing compliance with regulatory requirements

By setting up standardised processes, ISO 9001 helps organisations to comply with the legal standards applicable to their sector. This boosts stakeholder confidence and helps avoid potential sanctions.

5.3. ISO 31000: Risk management

ISO 31000 provides a framework for risk management in organisations:

Identifying risks

The standard encourages companies to conduct in-depth analyses of their internal and external environments. This may include sector assessments, market research and SWOT analyses to identify potential threats.

Risk assessment

ISO 31000 proposes risk assessment methods that enable organisations to determine not only the likelihood of risks occurring, but also the consequences that would follow. This helps to prioritise risks and to allocate the appropriate resources.

Risk management

Depending on the results of the assessment, companies can develop strategies to deal with the risks identified, whether by avoiding, reducing, transferring or accepting the risks. ISO 31000 also encourages the implementation of action plans to manage risks should they occur.

5.4. NI 43-101 (Canadian standard)

The NI 43-101 standard is essential for the mining sector in Canada:

Investor protection

By imposing rigorous requirements for the filing of technical reports, NI 43-101 ensures that information about mineral resources is accurate and based on verifiable studies. This ensures the reliability of the information provided to investors.

Ensuring transparency

The standard provides clear guidelines on how companies should disclose relevant information about their mining projects. This strengthens investor confidence and promotes better communication about the risks associated with projects.

Weaknesses :

• Perceived complexity: Small businesses often find the framework difficult to understand and apply, which can hamper their ability to raise finance.
• Reporting costs: The costs associated NI 43-101 compliance can be prohibitive for small businesses.
• Rigidity: The framework can innovation, by not enough room for the application of more flexible practices.

Divergences :

Strict qualifications: NI 43-101 requires experts to be specifically qualified, which can be an obstacle for new companies or less experienced practitioners.

5.5. JORC Code (Australian Standard)

The JORC Code, recognised in Australia, aims to guarantee the quality of information on mineral resources:

Guaranteeing the accuracy of information

The JORC Code stipulates that resource claims must be supported by reliable and verifiable data. This implies rigorous geological studies and systematic reporting of results.

Promoting integrity

By establishing strict reporting practices, the JORC Code enhances investor credibility and confidence, while ensuring that the information disclosed is fair and reasonably based.

Weaknesses :

• Subjective interpretation: The JORC code relies in part on the judgement of professionals, which can lead to varying ratios depending on individual interpretations.

• Slow updates: The process of revising and updating the code can be slow, making it less responsive to innovations and developments in the sector.
• Lack of international standardisation: Recognition and adoption of JORC outside Australia is limited, which can be a problem for companies seeking to raise funds internationally.

Divergences :

Focus on honesty: Unlike NI 43-101, which imposes strict formal requirements, JORC values transparency and integrity.

5.6. SAMREC Code (South African standard)

The SAMREC Code is essential mining reporting in South Africa:

Standardising reports

It establishes uniform criteria for the recognition and disclosure of mineral resources, facilitating clear and consistent communication between companies and stakeholders.

Setting quality standards

Compliance with this code guarantees that the information disclosed complies with rigorous standards of transparency and accuracy, thereby enhancing investor confidence.

Weaknesses :

• Limited international recognition: Although it is respected in South Africa, SAMREC is not widely applied internationally.
• Concentration on the mining sector: The framework is strongly focused on conventional resources, which limits its use for non-conventional resources such as rare earths or construction materials.

Divergences :

Integration of cultural and social elements: SAMREC takes into account the social and cultural impacts of mining projects, which is less frequently observed in other codes.

5.7. CRIRSCO

CRIRSCO plays a key role in harmonising reporting standards:

International coordination

By unifying international reporting standards, CRIRSCO makes it possible to compare the information provided by different jurisdictions, thereby facilitating decision-making for investors.

Improved transparency

By establishing common standards, CRIRSCO is encouraging greater transparency in the disclosure of mineral capacities and resources, thereby contributing to confidence in the sector.

Weaknesses

- General standards: The lack of strategic detail can lead to varying interpretations in different member countries.
- Variable implementation: Countries interpret and apply standards unevenly, which can lead to inconsistencies.

Divergences

International harmonisation: Although niche, this focus on harmonisation can have a limited impact on specific projects at local level.

5.8. SPEE

The SPEE is an authority on oil and gas resources:

Integrity of reporting

The SPEE ensures that reporting methods comply with industry best practice, thereby guaranteeing the reliability of the information provided.

Disclosure guidelines

The clear guidance offered by the SPEE helps companies to assess and disclose their reserves and resources, ensuring that the data shared is relevant and accurate.

5.9. AMSA

AMSA is responsible for safety standards in Australian mining industry:

Protection of workers

Thanks to rigorous regulations, AMSA ensures the well-being of workers, minimising the risk of accidents and improving working conditions.

Minimising risks

AMSA promotes robust safety practices and provides training to make workers aware of potential hazards, ensuring a safer working environment. Chapter 5 discusses the importance of international standards in the management of operations and transparency within industrial sectors, particularly those related to the environment and mineral resources.

These standards, such as ISO 14001 for environmental management, ISO 9001 for quality, and NI 43-101 mining expertise, provide a structured framework for organisations wishing to improve performance, ensure regulatory compliance and build trust with their stakeholders.

ISO 14001 promotes a proactive approach to environmental sustainability, encouraging companies to reduce their ecological footprint, improve resource use and develop effective waste management strategies.

This framework not only meets legal requirements, but also fosters a culture of continuous improvement that can enhance an organisation's reputation in the marketplace.

Similarly, ISO 9001 emphasises the importance of customer satisfaction and internal process improvement. By integrating quality management systems, companies can reinforce their commitment to operational excellence, while complying with legal requirements, which contributes to a significant competitive advantage. In the specific area of risk management, ISO 31000 encourages organisations to carry out systematic assessments of their environments in order to draw up robust action plans.

This standard promotes proactive risk management, enabling companies to navigate in often complex and uncertain environments.

In the mining sector, the Canadian NI 43-101, Australian JORC Code and South African SAMREC standards underline the importance of accurate and transparent disclosure of mineral resources.

Each of these standards, with their specific strengths and weaknesses, aims to investors and promote integrity in mining reporting. However, challenges such as high compliance costs, subjective interpretations and a lack of international recognition remain barriers to wider adoption.

CRIRSCO plays a facilitating role in harmonising mineral reporting standards internationally, while highlighting the differences that exist between the various standards. While SPEE and AMSA focus on integrity and safety in the oil and gas and mining sectors, respectively, they also illustrate the industry's overall commitment to sustainable and safe practices.

In short, these international standards not only help to establish guidelines for the conduct of business, they also enhance investor and consumer confidence, promote social and environmental responsibility, and contribute to sustainable development. As the world continues to face growing environmental and economic challenges, the integration of these standards is becoming essential to the sustainability and success of businesses worldwide. It is therefore crucial for organisations to commit to respecting and implementing these standards, while seeking to innovate in their practices to meet the needs of a changing world.

CHAPTER 6
NATIONAL STANDARDS

The regulatory framework is essential in the field of mining and the evaluation of subsoil and soil resources. Each country develops its own standards that govern practices to ensure safety, sustainability and environmental protection. In this chapter, we will explore several national standards, focusing on the Mining Code of the Democratic Republic of Congo, the Référentiel des sols in France, as well as BS 5930 and NF P 94-500.

6.1. Mining Code

The Mining Code is the cornerstone of mining regulation in many countries. This code defines the rules governing the exploration and exploitation of mineral resources. It is important for establishing a legal framework for the granting of operating licences, the associated rights and the obligations of operators. Environmental requirements, which aim to minimise the impact of mining activities on local ecosystems, are also incorporated into the legislation.

Main objectives of the Mining Code :

- Permits: Ensuring that mining operators have the necessary rights to explore and exploit resources. This includes checks on the technical and financial capacity of applicant companies, thus ensuring that only competent and responsible entities have access to resources.
- Environmental : Establishing preventive and corrective measures to limit environmental damage. This includes carrying out environmental impact studies before the start of activities, identify and mitigate potential risks to ecosystems.
- Community rights: Recognising and protecting the rights of local populations affected by mining activities. This means ensuring that they are consulted and that their concerns are taken into account in the decision-making process.

6.1.1. Mining Code of the Democratic Republic of Congo

The Mining Code of the Democratic Republic of Congo (DRC) is a set of laws and regulations governing the exploration, exploitation and management of mineral resources in the country, which is rich in minerals such as copper, cobalt, gold and diamonds. Mining legislation is crucial to economic development and sustainable resource management.

a. Historical background

Ordinance-law no. 81-013 of 02 April 1981, containing general legislation on mines and hydrocarbons, was repealed and replaced by law no. 007/2002 of 11 July 2002, containing the Mining Code, and then revised in March 2018. The revision aims to attract more foreign investment while ensuring that the benefits of mining accrue to the Congolese people. revision process involved consultations with various stakeholders in the sector, including NGOs and local communities.

b. Objectives of the Mining Code

- To supervise mining operations (industrial and artisanal): To ensure that all mining activities are carried out legally and responsibly, while complying with national laws and regulations.
- Protect the environment: Include provisions on environmental management, monitoring tools and rehabilitation plans for mining sites.
- Guaranteeing the rights of local communities: Taking into account the social impacts of mining and promoting local development through community projects.
- Encouraging investment: Providing a stable and predictable framework for investors, while protecting the interests of the State through periodic contract review clauses.

c. Main provisions

- Mining permits: The Code defines the different types of permits (exploration permits, operating permits, etc.) and the procedures for obtaining them, including rigorous selection criteria.
- Royalties and taxes: Establishment of a system of royalties and taxes for mining companies to ensure equitable revenue sharing, with percentages earmarked for community development programmes.
- Corporate social responsibility: Companies must carry out social and environmental impact

assessments and involve communities in the decision-making process, with regular reports on progress made.

• Control and monitoring: Establishment of bodies to monitor and regulate mining activities, including regular inspections and environmental audits.

d. Towards Sustainable Management of Mineral Resources

Congolese legislation on the management of mineral resources aims to guide and regulate mining and exploration. However, the immediate profitability of exploitation naturally attracts investors, who are looking for quick returns on their investments, while exploration, which involves high risks and substantial initial investment, is often perceived as a less attractive venture. This strong preference for exploitation can have a number of adverse consequences. Economically, it can lead to over-exploitation of resources, leading to a long-term scarcity of minerals. Environmentally, the focus on exploitation could exacerbate the problems of degradation of local ecosystems, pollution and other environmental impacts. Finally, in socio-political terms, such a dynamic could fuel conflicts over the distribution of resources and wealth, exacerbating local inequalities.To remedy this situation and balance investor between exploration and exploitation, the Congolese government needs to initiate attractive strategies:

Tax incentives

Put in place an attractive fiscal framework for investment in mineral exploration. This could include temporary tax exemptions reductions in exploration fees during the early stages of exploration.

Strengthening the infrastructure

Invest in the necessary infrastructure (roads, ports, etc.) in regions with high mining potential. Better access to exploration sites can reduce costs and risks associated with exploration.

Public-private partnerships

Encourage partnerships between the state and private companies (multinational and local) to share the costs and risks of exploration. This could also include state-owned companies playing an active role in exploration.

Access to information and technology

Facilitating access geological and geographical data, as well modern exploration technology, to help investors make informed decisions. This includes publishing reports and feasibility studies and regularly updating the country's geological and metallogenic map. Finally, it is essential that the Democratic Republic of Congo adopts a balanced approach to the management of its mineral resources. By implementing attractive strategies to stimulate exploration, the government will be able to not not only ensure the sustainable exploitation of resources, but also create an environment conducive to investment, economic growth and local development. Attention to these dynamics could make the DRC a model of responsible natural resource management, while bringing tangible benefits to its people.

6.2. Soil repository

In France, the Référentiel des sols plays a crucial role in the assessment of contaminated sites. This standard aims to ensure the appropriate management of polluted land, by providing criteria analysing soil quality. It identifies potential risks to human health and the environment, making it easier to decide on the necessary remediation measures.

Main features of the Soil Repository :

- Risk assessment: Identify the contaminants present and assess their potential impact on health and the environment, establishing alert thresholds and intervention levels.
- Analysis methodology: Provide protocols for characterising contaminated soils, including sampling and laboratory analysis methods.
- Site rehabilitation: Propose guidelines for land rehabilitation and the restoration of affected ecosystems, incorporating sustainable rehabilitation strategies.

6.3. BS 5930 (United Kingdom)

BS 5930 is a British Standard that establishes a framework of practice for geotechnical investigation. It provides detailed guidance on the planning and execution of geotechnical investigations, which are essential to ensure the safety and stability of buildings and infrastructure.

Key elements of BS 5930 :

- Survey planning: Advice on the design of investigations to meet specific project needs, including the choice of methods and the definition of study objectives.
- Investigation techniques: Description of the methods and tools to be used to assess soil conditions, such as drilling, in situ tests and laboratory analyses.
- Reporting and documentation: Guidance on how to document and communicate the results of geotechnical studies, emphasising the importance of clear and accurate communication with stakeholders.

6.4. NF P 94-500 (France)

Standard NF P 94-500 is an essential reference in France for study and analysis of geotechnical soils. It defines the methods to be used for field investigations and the assessment of geotechnical properties, guaranteeing the quality and reliability of the data collected.

Fundamental aspects of NF P 94-500 :

- Investigation methods: Draw up protocols for carrying out laboratory and in situ tests, specifying the equipment to be used and the procedures to be followed.
- Soil characterisation: Define the geotechnical parameters needed to assess the behaviour of soils under different conditions, including strength, compressibility and permeability analyses.
- Quality standards: Ensuring that investigation practices comply with safety and sustainability requirements, while incorporating quality criteria for the data collected and reports.

tandardisation in France is regulated by the law of 24 May 1941 and decree no. 2009-697 of 16 June 2009, which defines the tasks of the Association française de normalisation (AFNOR).

6.5. AFNOR

The Association Française de Normalisation (AFNOR) is the national standards body in France. It plays an essential role in the development and promotion of standards in various sectors and industries. AFNOR strives to ensure the quality, safety and interoperability of products and services, thereby improving consumer protection and facilitating international trade.

AFNOR's main functions include :

- Development of standards: AFNOR is involved in the creation of French and international standards in collaboration with industry stakeholders, government bodies and other organisations.
- Certification: The organisation offers certification services for products, systems and services to ensure their compliance with specific standards and regulatory requirements.
- Training and information: It offers training programmes and resources to help organisations understand and implement the standards effectively.
- Advocacy: It promotes the importance of standardisation in improving the competitiveness, innovation and sustainability of French businesses.
- International collaboration: It represents France in international standards bodies such as ISO (International Organization for Standardization) and IEC (International Electrotechnical Commission).

This chapter has highlighted the importance of the regulatory framework in the field of mining and evaluation of subsoil and soil resources. By analysing the DRC's Mining Code, France's Référentiel des sols, NF P 94-500, and the BS 5930 standards, we have shown how these regulations serve not only to provide a framework for mining activity and soil management, but also to guarantee sustainability, safety and environmental protection.

The DRC's Mining Code, while an essential tool for attracting investment, emphasises the need for a balance between resource exploitation and environmental and social concerns. This implies responsible management that takes into account the rights of local communities and ensures a fair share of the benefits.

France's Référentiel des sols provides a structured framework for assessing and remediating contaminated sites, which is fundamental to protecting public health and the environment. The BS 5930 and NF P 94-500 standards reinforce this approach, providing precise guidelines for geotechnical investigations and soil analysis, guaranteeing the reliability of the data required for safe and sustainable infrastructure projects.

An examination these standards makes it clear that standardisation and regulation are essential elements both for the management of natural resources and for protection of ecosystems and the communities that depend on them. By promoting responsible practices and fostering transparency in mining and agricultural operations, these regulatory frameworks have the potential ensure sustainable development while safeguarding the interests of future generations.It is therefore imperative that governments, businesses and communities work together to strengthen the application of these standards and develop new strategies that promote the balanced and sustainable use of natural resources. Proactive implementation of these regulations can turn challenges into opportunities. opportunities, enabling us to move towards a future in which natural resources are used responsibly, both for economic benefit and to preserve the environment and the rights of local communities.

CHAPTER 7

STANDARDS SPECIFIC TO MINING INDUSTRY

7.1. Guidelines Environment

Environmental guidelines are an essential requirement for any mining project. They aim to ensure that mining operations minimise their environmental footprint while complying with current legislation.

Environmental Impact Assessments (EIA)

Before any mining operation begins, a thorough EIA is essential. This process begins with data collection on the local environment, involving soil, water and wildlife surveys. The EIA examines potential impacts such as water pollution, deforestation and habitat disturbance. Consultations with stakeholders, including local communities, are essential to gather concerns and suggestions. Based on the findings of the EIA, mitigation measures must be developed, such as the creation of buffer zones around sensitive habitats, to reduce long-term impacts.

Waste Management

Mining operations generate various types of waste, from extracted waste rock to processing residues. Effective management this waste is crucial to prevent contamination of soils and . This means setting up systems that secure storage facilities for hazardous products and the implementation of technologies waste treatment systems that minimise solid and liquid waste. Companies must also comply with local and international regulations waste treatment, and develop contingency plans in the event of a spill or contamination.

Restoration of Mining Sites

Site restoration is a key responsibility of mining companies. Guidelines should stipulate that operations must include rehabilitation plans from the outset, aimed at restoring the local ecosystem after mine closure This may include replanting native vegetation, revitalising soils and rehabilitating watercourses. Companies should also establish partnerships with environmental organisations to ensure that their restoration efforts are effective and sustainable.

7.2. Health and Safety Management Standards

The safety of workers in the mining sector is of paramount importance, given the hazardous nature of operations.

Risk Assessment

Every mining site must carry out regular risk assessments to identify potential hazards. This involves carrying out scenario analyses for landslides, explosions and chemical hazards. The results of these assessments must lead to development of safety protocols aimed at minimise workers' exposure to these hazards. Practical exercises must also be carried out regularly to prepare staff to react effectively in the event of an emergency.

Worker training and awareness

Adequate training is essential to ensure that every employee is aware of the hazards and safety protocols in place. Companies should ongoing training that cover the correct use of personal protective equipment (PPE), evacuation procedures, and simulated evacuations in the event of a fire or major accident. In addition, awareness of the mental health and well-being of workers must be incorporated into these training courses.

Worker Health Monitoring

Regular monitoring programmes are needed to detect health problems linked to working conditions at an early stage. This includes periodic medical examinations to monitor exposure to hazardous substances, as well as health awareness campaigns to inform workers of the dangers associated their environment. Companies should also establish confidential communication channels enabling employees to report health problems without fear of reprisals.

7.3. GIA (Guidelines for the Internationalization of Geoscience)

GIAs are an essential framework for harmonising geoscientific practices across the world.

Best Practices for Research and Exploration

GIAs provide recommendations on research and exploration methods, encouraging the use of scientifically validated techniques for collection and analysis of geoscientific data. This includes the use of advanced technologies such as remote sensing and geological imaging to improve the accuracy of geological assessments. Companies must also ensure that they document and share their research results to contribute to the global knowledge base.

International collaboration

GIAs encourage collaboration between geoscientists from different countries to share knowledge and technologies. This can take the form of joint research projects, international conferences and information-sharing platforms. International cooperation helps to harmonise standards and build technical capacity in developing countries, thereby promoting responsible and efficient mining practices worldwide.

Integrating social considerations

The GIAs stress the importance of considering the societal impacts of geoscience projects. This means engaging with local communities to understand their concerns and incorporate their perspectives into mining projects. Companies should promote initiatives aimed at fostering sustainable development, creating local jobs and supporting community projects.

7.4. Minerals Council of Australia

The Minerals Council of Australia (MCA) plays an important role in promoting sustainability within the mining industry.

Corporate Social Responsibility (CSR)

The MCA promotes the adoption CSR practices in the mining sector, encouraging companies to consider not only their profits, but also their impact on local communities and the environment. This includes transparency in operations and communication of financial and environmental results to stakeholders. Companies must establish partnerships with local communities to support economic and social development.

Sustainable development

The MCA framework sets out clear standards for sustainable development, emphasising the importance of managing environmental, social and economic impacts. This includes the implementation of strategies to reduce greenhouse gas emissions. and minimise water consumption. Companies should also consider investing in clean technologies and circular economy practices to improve their sustainability.

Innovation and Technology

The MCA encourages the adoption of innovative technologies to improve the efficiency of mining operations. This includes the use of automation to reduce the need for labour in hazardous environments, as well as the development of recycling techniques to reuse extracted resources. Companies need to invest in research and development to explore new mining methods that minimise environmental impact.

Chapter 7 highlighted the importance of specific standards governing the mining industry, highlighting both environmental guidelines and health and safety management standards, as well as the impact of initiatives such as GIA and the efforts of the Minerals Council of Australia.

Environmental guidelines, such as environmental impact assessments, waste management and mine site restoration, play a fundamental role in ensuring that mining projects operate in a responsible and sustainable manner. These processes not only minimise the ecological footprint, but also encourage the rehabilitation of local ecosystems, showing that industry can co-exist with the natural environment. Consultation with local communities is essential to integrate their concerns and aspirations into projects, promoting equitable and inclusive development.

The health and safety of workers is also at the heart of industry standards. Risk assessment, ongoing training and employee health monitoring ensure a safer working environment, reducing incidents and improving worker wellbeing. The focus on mental health and worker awareness underlines a holistic approach to safety, which is essential in a sector that is often exposed to hazardous conditions.

The GIAs represent a significant step towards harmonising geoscientific practices at international level, encouraging best practice in research and exploration. International collaboration and the integration of societal concerns are essential elements ensuring that mining projects take account of local realities while promoting standards of excellence.

Finally, the Minerals Council of Australia embodies the commitment to a sustainable mining industry. By integrating corporate social responsibility, sustainable development and technological innovation, the MCA establishes a framework that enables companies not only to maximise their profitability, but also to contribute to a more sustainable mining industry. communities and the environment. In short, this chapter demonstrates that the mining industry, when guided by strict standards and responsible practices, can move towards a sustainable and ethical path. By adhering to these standards, mining companies can not only succeed economically, but also play a leading role in preserving the environment and the well-being of communities, setting a model for the future of mining worldwide.

CHAPTER 8

OTHER STANDARDS

In the field of geological engineering and natural resources, standards play a fundamental role in guaranteeing the quality, safety and sustainability of operations. Two major organisations, ASTM International and the American Petroleum Institute (API), set standards that have a major influence on the practices and technologies employed in these sectors.

8.1 Standards ASTM

Founded in 1898, ASTM International, formerly the American Society for Testing and Materials, is a not-for-profit standards organization. Its objective is to develop consensus standards for a wide range of materials, products, systems and services, making it a key player global quality assurance. ASTM standards are particularly essential in a variety of sectors such as construction, energy, the environment and mining. Developed by experts and professionals, these standards reflect the latest technological advances and best practices in the field. In geological engineering, ASTM standards provide rigorous test methods for evaluating various materials such as concrete, soils and rocks. These methods include the compressive strength test, which assesses a material's ability to withstand axial loads, thus ensuring the strength of structures. The permeability test measures the capacity a soil or rock to allow water to pass through it. This information is crucial for managing water resources and preventing pollution. Finally, chemical analyses provide an assessment of the composition of materials, essential for identifying geotechnical and environmental properties. The application of ASTM standards also ensures the comparability of test results between different laboratories. This uniformity is essential in critical contexts where decisions such as foundation design or natural hazard assessment are based on geological data. It establishes a common language among professionals, making it easier to understand and interpret results.

8.2 Standards API

At the same time, the standards of the American Petroleum Institute (API), created in 1919, govern practices and technologies within the oil and gas industry. These standards cover all aspects of the industry, from exploration to the production and distribution of oil and gas. API standards are fundamental to ensuring safe and sustainable operations. They include specifications for the geological and geophysical evaluation of reservoirs, helping to define reservoir configurations and properties, which are crucial for efficient extraction. API standards also address pipeline integrity by defining methods to minimise the risk of leaks and failures, events that could lead to disastrous environmental consequences. Extraction practices, , are covered by specifications on drilling and well completion, ensuring that the methods employed comply with best practice in terms of safety and sustainability.

Compliance with these standards enables companies not only to ensure the safety of their operations, but also to comply with government regulations. They also provide a framework for technological innovation, encouraging the adoption of new methods to improve efficiency and reduce the risks associated with oil and gas operations.

8.3 Geologists' Protocol and Assessment of Resources

Protocols for the evaluation of geological resources are essential documents that establish precise criteria for the quantification and classification of geological resources, including ores, hydrocarbons and construction materials. These protocols follow internationally recognised standards, defining the requirements for the disclosure of information on mineral resources.

8.4 Approach Systematic

Resource assessments require a systematic approach, involving the collection of geological, geochemical and geophysical data. The protocols specify sampling methods, analysis techniques and resource estimation models, ensuring that estimates are made objectively and transparently. This is crucial for decision-making on investment and the development of mining projects.

8.5 Documentation and Transparency

In addition, these protocols require full documentation of results, enabling critical peer review and guaranteeing data traceability. By adhering to these standards, geologists can provide reliable reports that support the economic viability of projects. This contributes to the responsible management of natural resources, ensuring that decisions are based on accurate and verifiable data.In geological engineering, standards play a crucial role in guaranteeing the quality, safety and sustainability of industrial practices. ASTM and API standards, in particular, are fundamental pillars for the assessment and exploitation of geological resources, contributing to both environmental protection and operational safety.

ASTM standards, with their rigorous test methods, provide a sound scientific basis for the analysis of materials such as soils, rocks and concrete. They enable geological engineers to make informed decisions. for the design and installation of structures, ensuring that they meet strength and durability requirements. By promoting comparability of results between different laboratories, ASTM standards also encourage effective collaboration between professionals in the sector, which is essential in a field where critical decisions are based on accurate data.

At the same time, API standards, with their extensive reach across the oil and gas industry, ensure safe operations while stimulating technological innovation. By defining criteria for the geological evaluation of reservoirs and the integrity of pipelines, they minimise environmental risks and reinforce corporate responsibility. By incorporating these standards into their practices, industry players not only comply with regulations, but also contribute to the move towards more sustainable and efficient extraction methods.

Finally, the protocols established for the evaluation of geological resources ensure that resource estimates are carried out systematically and transparently. This is essential for strategic decision-making in developing mining projects and attracting investment. By ensuring rigorous documentation and promoting an objective approach, these protocols strengthen stakeholder confidence and support the responsible management of natural resources. In short, adherence to these standards and protocols is essential to establish a working framework that prioritises safety, sustainability and transparency. This not only helps to meet the current challenges of geological engineering and the resources industry, but also paves the way for future resource exploitation that respects both economic and environmental requirements. Standards, such as those established by ASTM and API, as well as protocols for geologists, are therefore essential foundations on which the future this vital sector rests.

CHAPTER 9

HARMONISATION OF DECLARATION STANDARDS FOR MINERAL RESOURCES AND OUTLOOK FOR THE DRC

The importance of norms and standards in the declaration of mineral resources cannot be underestimated. They play a vital role in promoting transparency, credibility and investor confidence in the mining sector, which often faces considerable challenges. In a global context where the demand for mineral resources is constantly increasing, it is essential for resource-rich countries such as the Democratic Republic of Congo to also equip themselves with regulatory frameworks.

The DRC, with its vast mineral reserves, is one of the most promising countries mining. However, the absence of clear and consistent standards for the exploration and evaluation of mineral resources leads to disparities and a lack of confidence among investors. This situation is exacerbated by structural weaknesses such as inconsistent standards, limited transparency and insufficient local training capacity. In a country where resources are both an opportunity for development and a potential source of conflict, it is imperative to establish a regulatory framework that can guarantee not only the accurate valuation of resources, but also the responsible and sustainable governance of these resources. This chapter looks at international mineral resource reporting standards, including JORC, NI 43-101, SAMREC and CRIRSCO, and examines the current situation in the DRC, the challenges it faces, and offers some suggestions for improving the situation. By adopting international best practices while adapting them to local realities, the DRC can not only enhance the value of its mineral resources, but also strengthen investor confidence, promote transparency and ensure sustainable development for its citizens.

9.1. Presentation of Norms and Standards

Standards and guidelines for mineral resource assessment, such as JORC, NI 43-101, SAMREC and CRIRSCO, play a vital role in establishing transparency and confidence within the mining industry. However, improvements can be made to increase their effectiveness and global adoption. Firstly, it is essential to establish national guidelines aimed at reducing subjectivity in the resource evaluation process. Variability in interpretations and approaches can lead to inconsistencies, both in reporting and in investor confidence. Training workshops should be organised to raise awareness among stakeholders, including geologists, engineers and analysts, of best practice and existing international standards. These workshops would promote a common understanding and consistent application of the guidelines, helping to strengthen the credibility of assessments. Secondly, simplifying assessment processes and disclosure requirements according to project size is another key improvement to consider. Currently, the requirements can appear overly burdensome for smaller projects, discouraging the exploration and development of new deposits. By introducing a proportionate framework, smaller companies and emerging projects could benefit from a more agile approach, allowing them to focus on exploration and innovation without being overwhelmed by disproportionate administrative obligations.

By incorporating sufficient flexibility to assess a variety of resources, it would be possible to adapt the methodologies to the specific geological and economic circumstances of each project. This would not only encourage innovation, but would also make it possible to better reflect the reality of projects in the field, while maintaining a rigorous evaluation framework.

The development of standardised implementation tools and the promotion of case studies illustrating successful applications are essential adherence to the standards. The creation of practical guides and dedicated software could facilitate the application of the directives while providing a frame of reference for practitioners. Case studies, meanwhile, would serve as inspiring models, demonstrating how companies have successfully navigated the assessment process while complying with established standards. This would encourage others to follow best practice and engage in responsible and transparent valuations. In short, by combining these improvements, clear guidelines, obligations, flexible international recognition and standardised implementation tools, we can not only strengthen the validity of resource assessments, but also stimulate innovation and sustainability within mining industry. These concerted efforts will contribute to a future where transparency and trust are at the heart of exploration and mining activities.

9.2. Situation in the Democratic Republic of Congo

The Democratic Republic of Congo has a wealth of mineral resources, which represent considerable economic potential. However, the country faces major challenges managing these resources. The absence of clear and consistent standards for the exploration and evaluation of mineral resources complicates the government's task. This situation creates difficulties in the efficient management of resources, creating a climate of uncertainty for investors.

9.2.1. Weaknesses Current

The weaknesses in the current system can be summarised in several points. Firstly, an inconsistency in the adoption of standards makes it difficult to understand and apply good practice. Investors, who lack confidence, may be discouraged by a lack of very reliable information about mining projects, which could also damage the country's reputation on the world market. In addition, there is insufficient local capacity to raise awareness of international standardswhich limits ownership of the standards. required by economic operators.

9.2.2. Suggestions for improving the situation in the DRC

To overcome these weaknesses, a number of suggestions can be implemented to create a robust regulatory framework and encourage the sustainable development of mineral resources.

a. Establishing a Local Standards Framework

It is essential to establish a solid local standards framework based on international best practice. The DRC could adopt standards such as JORC, NI 43-101, SAMREC and CRIRSCO to develop a national assessment code. This approach would require adaptation to local realities, taking into account the country's diverse geological, environmental, economic and social contexts. In addition, it would be necessary involve all stakeholders, including government, business, NGOs and local communities, in the development of this framework to ensure its effectiveness and acceptance.

b. Local Capacity Building

Building local capacity is another cornerstone of this strategy. The organisation of professional training courses on international standards for professionals in the mining sector is essential. In addition, partnerships with academic institutions could be initiated to integrate these standards into training curricula. It would also be beneficial to raise awareness among local communities about challenges of transparency and resource exploitation, so that they can play an active role in the process.

c. Implementation of a Control and Reporting System

The creation of a monitoring and reporting system is also essential. A dedicated monitoring agency could be set up to supervise mining activities and ensure compliance with established standards. It is essential to implement mandatory reporting mechanisms for companies, encouraging them to submit regular reports on their mineral exploration and evaluation activities. Data transparency must be a priority, with efforts to make certain reports and company performance public.

d. Promoting transparency

Promoting transparency is essential to boost investor confidence. The creation of a public information platform could make it possible to publish the results of environmental and community impact assessments and studies. , the establishment of sustainability indicators would help to measure and monitor the impact of mining projects. The commitment of international partners will also be crucial in collaborating establish transparency standards and their implementation.

e. Working with investors

Finally, collaboration with investors should be encouraged. This could include tax incentives for companies that comply with the standards, as well as creation of a certification label for compliant companies to recognise their efforts. Promoting long-term partnerships that incorporate sustainability principles will ensure a framework conducive to responsible investment and economic growth. We close this chapter on the harmonisation of Mineral Resource Declaration Standards and Outlook for the DRC with the following: Harmonising standards for declaring mineral resources is a key issue for the Democratic Republic of Congo, a country rich in resources but facing significant challenges in terms of transparency and investor confidence. This chapter has highlighted the importance of establishing a robust standards framework that incorporates international best practice while taking account of local specificities. Standards such as JORC, NI 43-101, SAMREC and CRIRSCO offer valuable models that, once adapted, could make a significant contribution to the development of the mining sector in the DRC. We have identified a number of weaknesses in the current system, including inconsistent standards, a lack of awareness and insufficient local capacity. If left unaddressed, these issues will continue to damage the DRC's reputation on the global market and slow down the investment needed to turn its mining potential into an economic reality. In response to these challenges, concrete suggestions were put forward, ranging from the establishment of a local normative framework to the promotion of transparency and collaboration with the investors. In conclusion, the Democratic Republic of Congo has the opportunity establish a regulatory framework aimed at developing its mineral resources while promoting sustainable and equitable development for its population. By strengthening local capacities, putting in place control and reporting systems, and promoting greater transparency, the country will be able to create an environment conducive to responsible investment. This will enable it to transform its natural resources into levers for economic development, while ensuring enlightened governance that benefits society as a whole. The commitment of stakeholders, including government, business and civil society, will be crucial to achieving this vision and building a future where the country's mineral wealth is managed in a way that is sustainable and beneficial to all. This issue will be explored in depth in the next chapter.

CHAPTER 10

TOWARDS THE ESTABLISHMENT A CONGOLESE STANDARDS FRAMEWORK FOR MINING REPORTING

The Democratic Republic of Congo is undeniably one of the richest countries in the world in terms of mineral resources. Vast deposits of minerals such as cobalt, copper, gold, lithium, Colombo-Tantalite and diamonds not only represent considerable economic potential, but also offer a unique opportunity for sustainable development for the country. However, to take full advantage of these resources, it is vital to establish a regulatory framework for the presentation and filing of technical reports, ensuring that information on mineral resources is accurate, based on verifiable studies and of high quality. This will ensure the reliability of the information provided to investors. It is in this context that a Norme Congolaise pour le Reporting Minier (NCRM) could play an essential role.

10.1. Objectives of the NCRM

Protecting investors

Investor protection is necessary to attract foreign capital to the Congolese mining sector. This means putting in place a framework that guarantees the reliability and verifiability of information on mineral resources.

Ensuring Transparency and Integrity

Ensuring transparency and integrity is essential to creating an environment favourable investment climate. It is therefore necessary to promote clear and precise communication of information, while taking account of socio-cultural and environmental issues.

Promoting sustainable development

Incorporating socio-cultural and environmental issues into reporting helps to ensure that mining projects do not harm the environment or local communities. Case studies of environmentally friendly mining projects, which have managed to balance economic profitability with the preservation of ecosystems, could serve as a benchmark. The NCRM could encourage companies to adopt sustainable natural resource management practices.

10.2. Guiding principles

Clarity and simplicity

For the standard to be effective, it must be easy to understand, even for small companies that do not always have the resources of a large multinational. Therefore, the creation of practical guides and simplified reporting templates could help these companies to comply with the requirements of the standard without incurring excessive costs.

Flexibility and Innovation

The standard should enable companies to adapt to technological developments, such as the use of mineral data management software or platforms. digital reporting.

Harmonisation

The NCRM must align itself with international standards. This will make it easier to compare Congolese projects with those in other countries, attracting international investors looking for transparency and rigorous reporting.

10.3. Key elements of the standard

Expert qualifications

Establishing qualification criteria for the experts involved in reporting is essential. This could include specific certifications for geologists, mining engineers and environmental specialists. To this end, the creation of a national register of certified experts would ensure that only qualified and competent people carry out studies and reports. It is in this context that professional bodies, in particular the Ordre National des Géologues de la RDC (OGRDC), will play a key role.

Standardised reporting

The adoption of uniform reporting formats would make it easier to compare projects and ensure consistency in the presentation of information, so an online platform where companies submit standardised reports could simplify the analysis of data by investors.

Performance Indicators

Developing clear indicators, for example sustainability indicators such as CO2 emissions per tonne of ore mined or the rate of remediation of mining sites, would make it possible to assess the performance of projects not only economically but also socially and environmentally.

10.4. Implementation mechanisms

Training and awareness-raising

Workshops and training seminars could be organised to raise awareness of NCRM requirements among companies and experts. As a result, training programmes in collaboration local universities could be set up to train the next generation of professionals in the mining sector.

Monitoring and assessment

The creation of an independent regulatory body could monitor the application of the standard. This regulator could also tasked with conducting regular audits and publishing reports on mining companies' compliance with the standards.

Reporting platform

The development of a dedicated digital platform would make it possible to harmonise the submission of reports and guarantee their accessibility. Investors would then be able to easily consult the information submitted by companies, thereby promoting transparency and accountability.

10.5. Integration of local issues

Socio-cultural impact

Socio-cultural impact assessments must take into account the traditions and values of local communities. Involving community leaders in the assessment process could help identify the potential impacts of mining projects on local culture and develop appropriate mitigation strategies.

Community involvement

Encouraging companies to engage in dialogue with local communities is essential. Initiatives such as community advisory boards or partnership programmes could build trust and enable companies to better understand the concerns of local people.

10.6. Review and update of the standard

Continuous Improvement Process

The introduction of a formal process for revising the standard every 3 to 5 years, based on feedback from users and developments in the sector, will ensure that the standard remains relevant. In conclusion, the introduction of the Norme Congolaise pour le Reporting Minier (NCRM) is a salutary initiative for the Democratic Republic of Congo, a country brimming with mineral wealth. By establishing a regulatory framework for the reporting and transparency of information on mineral resources, the NCRM aims to strengthen investor confidence while guaranteeing ethical and sustainable management of the mining sector.
In this context, the Ordre National des Géologues de la République Démocratique de Congo can play a very important role implementing this standard. As a regulatory and professional body, the Ordre has already taken a number of steps in line with this vision. Its contribution will be essential in designing the standard and ensuring that the qualification criteria for experts, performance indicators and standardised reporting are respected, thereby enhancing the credibility of the data. By integrating local issues and encouraging the involvement of communities in the decision-making process, the NCRM can initiate constructive dialogue

between the government, companies and affected populations, thereby promoting respectful mining practices. In addition, the process of continuous improvement of the standard will ensure that it remains relevant in the face of changes in the sector. In this way, the Norme Congolaise de Reporting Minier (NCRM) will represent not only a vector of confidence for investors, but also a lever for promoting sustainable and responsible mining development in the DRC. By adopting international best practice, the NCRM has the potential to become a model and a benchmark for other countries in the region, thereby contributing to the ethical and sustainable management of mineral resources.

CONCLUSION

This book has set out to explore the subject of standards used in geology and the mining industry, highlighting their key importance in regulating and structuring natural resource exploration and exploitation practices. Through various chapters, we have examined the historical evolution of standards, their scope of application, as well as the specific standards that govern the sector. We have also discussed international and national standards, the issues surrounding the transparency of mineral resources, and initiatives aimed at establishing regulations in the Democratic Republic of Congo, namely the Norme Congolaise de Reporting Minier (NCRM).The first major finding is that standards, whether social, environmental or technical, play a fundamental role establishing an ethical and responsible framework for mining. Not only do they guarantee the safety of workers and surrounding communities, but they also contribute to the sustainability of ecosystems. This is particularly relevant in a global context marked by growing concern for the environment and sustainable development. Companies that rigorously adopt these standards can enhance their reputation, attract investment and contribute to equitable economic development. With regard to the harmonisation of mineral resource declaration standards, as we discussed in the chapter on the Democratic Republic of Congo, it is imperative to establish a local standards framework that incorporates international best practice while taking into account the country's specific socio-cultural, environmental and economic characteristics. The establishment of the NCRM represents a significant step towards this goal, providing clear guidelines for transparency, social responsibility and resource management. This could not only boost investor confidence, but also improve the governance and management of mineral wealth for the benefit of the local population. On the other hand, challenges persist. Inconsistent standards, lack of awareness and insufficient local capacity are all obstacles that need to be overcome to ensure effective implementation of established standards. This requires the collaborative commitment of all stakeholders, including government, business and civil society. Awareness-raising and training are essential to develop local capacity and ensure a thorough understanding of standards requirements. In sum, this book highlights the importance of an integrated and holistic approach to standards in mining industry. Standards should not be seen solely as regulatory requirements, but as strategic elements that, when applied judiciously, can lead to sustainability, social responsibility and efficient and equitable resource exploitation. mining. The road to mining that is sustainable, ethical and beneficial to all stakeholders is of pitfalls, but the rigorous implementation of standards represents a promising route to achieving these goals.Finally, at a time when the world is facing major environmental and social challenges, the commitment to respecting and strengthening standards in mining industry not only a necessity, but also an opportunity to build a future in which the exploitation of natural resources can make a real contribution to the well-being of societies and the preservation of our planet.

BIBLIOGRAPHY

Academic articles and reviews

- Bisschop, L., & Kessler, A. (2016). International Mining Standards: An Overview of the Major Reporting Codes. Journal of Cleaner Production, 135, 1212-1221.

- Dinçer, I., & Tüzün, M. (2012). The role of standards in the mining industry. Resources Policy, 37(4), 215-222.

- Deux, Béatrice, et al (2013). Les normes : Enseignements et pratiques. Presses Universitaires de France.
- Gavrilovic, M., et al. (2018). The Role of Standards in Sustainable Mining Practices: A Critical Review. Resources Policy, 59, 60-69.
- Karl D. M. R. H. de L. V. Huberts, et al (2011). Mining and the Environment: From Ore to Metal. Mineral Resources Review.
- Michaud, V., & Davidson, A. (2016). The Future of Mining: Sustainable Practices in the 21st Century. Mining Engineering, 68(2), 31-37.
- Rhett, J., & Smith, J. (2018). The impact of ISO 14001 on mining operations: A case study. Journal of Cleaner Production, 198, 135-144.
- Schmidt, S. P. C., et al (2018). Regulating Mining in Africa: For People, the Planet and Profit. Open Society Foundations.
- W. K. B. G. Simon, et al (2020). Mining, Environment and Society. Routledge.

Norms and standards

- American Petroleum Institute (API). (2022). API Standards.

- ASTM International. (2023). ASTM Standards.

- Canadian Securities Administrators (CSA). (2011). National Instrument 43-101: Standards of Disclosure for Mineral Projects.
- International Organization for Standardization (ISO). (2015). ISO 9001: 2015 - Quality management systems - Requirements.
- International Organization for Standardization (ISO). (2015). ISO 14001: 2015 - Environmental management systems - Requirements and guidance for use.

• International Organization for Standardization (ISO). (2018). ISO 31000: 2018 - Risk management - Principles and guidelines.
• JORC. (2012). JORC Code - Australasian Code for Reporting of Exploration Results, Mineral Resources and Ore Reserves.
• South African Mineral Resource Committee (SAMREC). (2007). SAMREC Code for Reporting of Mineral Resources and Mineral Reserves.

Specialist books

• El Khoury, Rania (2018). ISO standards and reflections on standardisation. Various publishers.

• International Labour Organization (ILO). (2017). Safety and Health in Mining: A Guide for Managers. ILO Publications.

• Jansen, J. (2020). Geological Assessment and Resource Management in Mining. Wiley.
• Klohn, R. (2018).Environmental Management in the Mining Sector: Theory and Practice. Routledge.

• MacDonald, A. (2016). Mining and the Environment: From Ore to Metal. Cambridge University Press.

• Poudel, S., & Awasthi, A. (2020). Corporate Social Responsibility in the Mining Industry: A Global Perspective. T&F Books.
• World Bank Group (2019). Mining Regulations: Best Practices for Sustainable Operations. World Bank Publications.

Official documents

• Mining Code of the Democratic Republic of Congo (2018). Law n° 18/001 of 9 March 2018.
• Environmental Protection Agency (EPA). (2020). Developing and Implementing an Environmental Management System.

• David Hunter, et al. (2020). International Environmental Law and . Foundation Press.

• Boulanger, Pascal. (2017). Standardisation and quality management. Editions EMS.

• IUCN. Mining and Biodiversity: A Review of the Challenges and Opportunities.

• United Nations Environment Programme (UNEP). (2021). Environmental Guidelines for Mining Operations.

• World Health Organization (WHO). (2018).Occupational Health and Safety in the Mining Industry: A Guide for Managers.

Other works

• A. R. Partidário, et al (2000) Introduction to Environmental Impact Assessment. E & FN Spon.

• E. L. F. Wilkins (2015). The Mining Law of 1872: A Legacy of Ambivalence. Harvard Law Review.

• Minerals Council of Australia (2017). Enduring Value: A Framework for Sustainable Mining.

• World Bank. The World Bank Group & the Environmental and Social Framework.

Webography

Standards and organisations

• ISO - International Organization for Standardization. www.iso.org https://www.iso.org;

• Canadian Securities Administrators. www.securities-administrators.ca https://www.securities-administrators.ca

• JORC - Joint Ore Reserves Committee. www.jorc.org
http://www.jorc.org

• SAMREC - South African Mineral Resource Committee. www.samcode.co.za
http://www.samcode.co.za

• CRIRSCO - Committee for Mineral Reserves International Reporting Standards.
www.crirsco.com

http://www.crirsco.com

• SPEE - Society of Petroleum Engineers Experts. www.spe.org
https://www.spe.org

Additional resources

• World Bank - Mining Industry Overviewwww.worldbank.org/en/topic/mining

https://www.worldbank.org/en/topic/mining

• Minerals Council of Australia. www.minerals.org.au
https://www.minerals.org.au

• ASTM International. www.astm.org
https://www.astm.org

• American Petroleum Institute (API). zwww.api.org
https://www.api.org

• Mining Code of the Democratic Republic of Congo,
Loi N°18/001du 09/03/2018 portant code minier
www.leganet.cd/Legislation/Droit%20economique/Code%20Miner/Loi.18.
001.09.03.2018.html.

MIX
Papier aus verantwortungsvollen Quellen
Paper from responsible sources
FSC® C105338

Printed by Books on Demand GmbH, Norderstedt / Germany